Quantum Mechanics is incomplete and not paradoxical

CARLO MARIA PACE

QUANTUM MECHANICS IS INCOMPLETE AND NOT PARADOXICAL

Titolo | Quantum Mechanics is incomplete and not paradoxical
Autore | Carlo Maria Pace

ISBN | 978-88-92618-22-0

Youcanprint Self-Publishing
Via Roma, 73 - 73039 Tricase (LE) - Italy
www.youcanprint.it
info@youcanprint.it
Facebook: facebook.com/youcanprint.it
Twitter: twitter.com/youcanprintit

GENERAL INTRODUCTION

In this treatment I, by starting from the fundamental principles of Quantum Mechanics, demonstrate, in a scientifically rigorous way, that Quantum Mechanics is incomplete and not paradoxical.

In particular, in the first chapter, I demonstrate, contrary to the Copenhagen Interpretation (and to other similar interpretations) of Quantum Mechanics, that two physical quantities, which are described by operators that do not commute between them (relatively to a same physical system), exist, in any case, simultaneously relatively to a same observable physical system, even in the relativistic case. In other words, I demonstrate that Quantum Mechanics is incomplete, in the sense that the quantum wave function of a physical system does not describe completely the physical state of the physical system. Moreover, I demonstrate that my position is also supported by the correct interpretations of the (Heisenberg's) Uncertainty Principle, of the famous double-slit experiment, of the Einstein-Podolsky-Rosen Paradox and of the Bell's Theorem.

Furthermore, in the second chapter, I extend this to the field of application of the second quantization, in particular to the interpretation of the Feynman diagrams: the conservation of energy is valid also in the interactions that are described by Feynman diagrams, even below the indeterminacy of the energy in the formulation of the (Heisenberg's) Uncertainty Principle.

On the other hand, in the third chapter, I underline that my interpretation of Quantum Mechanics is supported also by the possibility (which is widely recognized) of using the Theory of De Broglie-Bohm in the field of application of Quantum Mechanics, even though my interpretation of Quantum

Mechanics is not equal to that of this theory and can be seen as an improvement of the Theory of De Broglie-Bohm.

Finally, in the general conclusion, I also disprove the paradoxical presence in Quantum Mechanics of physical effects without any cause: in other words, every effect has a cause, even in Quantum Mechanics.

CHAPTER I
THE CORRECT INTERPRETATION
OF THE INDETERMINACY
OF THE MEASURES
IN QUANTUM MECHANICS

1 The principal aim of this treatment

The principal aim of this treatment is to demonstrate, by starting from Quantum Mechanics which includes the (Heisenberg's) Uncertainty Principle, the falsity of the interpretation of Quantum Mechanics, which interpretation is unfortunately given by very numerous publications[1], where in accordance with this interpretation we would have that each physical state of every observable physical system in Quantum Mechanics would be completely described, from every physical point of view, by a quantum wave function and, moreover, that, however we choose two physical quantities which are described by operators that do not commute between them (relatively to a same physical system), these two physical quantities would not exist simultaneously relatively to a same observable physical system, however we have chosen this observable physical system. In particular, this erroneous interpretation is asserted not only by the so-called Copenhagen Interpretation (of Quantum Mechanics), but also by other interpretations of Quantum Mechanics, as the Many-Worlds Interpretation.

---

[1] Cf. LANDAU – LIFSHITZ pp. 15-21; HAWKING pp. 72-77; EISBERG – RESNICK pp. 65-68.77-80; BRANSDEN – JOACHAIN p. 58.

2 The non-relativistic case

As for the field of application of Non-Relativistic Quantum Mechanics, we begin by considering the component on the x-axis of the vector of position, that is to say the abscissa x, and the component on the x-axis of the vector of momentum, that is to say p_x, as physical quantities, which are described by operators that do not commute between them (relatively to a same physical system). (Analogously we could reason with regard to y and p_y or to z and p_z).

We know that in Quantum Mechanics one can consider a free particle (that is, a physical system, which is constituted by a unique particle that is not subject to any force, of which system the Hamiltonian is $H = \dfrac{p^2}{2m}$) with a p_x, which is known with an arbitrarily small, at the limit null, error and which is constant in time. Of this free particle we can measure the abscissa x, which is reached at the time t_0, without any lower limitation to the error of measure of the same abscissa x, in accordance with what the postulates of Quantum Mechanics say.

Now, for measuring the abscissa x of the particle at the time t_0, we, by definition, have to make the particle evolve in a free way until the time t_0, and then immediately we have to measure the abscissa x of the particle at the time t_0. Therefore the particle at the time t_0 has the abscissa x, which we measure with an arbitrarily small, at the limit null, error, and the p_x, which we already know with an arbitrarily small, at the limit null, error, because any free particle always keeps the same p_x and the particle is left free until the instant t_0. Thus, after the measurement of the abscissa x of the particle we know both $p_x(t_0)$ and $x(t_0)$ of the particle, that is to say x and p_x of the same particle at the same instant t_0 with arbitrarily small, at the limit null, errors of measure, without violating Quantum Mechanics.

But, on the other hand, the measurement of the abscissa x of the particle at the time t_0 perturbs the p_x of the same particle, and if we have measured the abscissa x (of the particle) at the limit with null error we have that the p_x (of the same particle) can be changed in any way. In the moment in which we have ended the measurement of the abscissa x of the particle we know $x(t_0)$ and $p_x(t_0)$ of the particle with, at the limit, null errors, but the p_x (of the particle) at the end of the measurement of the abscissa x (of the particle) is completely unknown because of the perturbation that the measurement of the abscissa x of the particle produces on the p_x of the same particle. At the end of the measurement, at the limit without error, of the abscissa x of the particle we practically have information relatively to the present physical state of the physical system only on the abscissa x of the particle (admitting that this abscissa has not changed in the meantime during the time of the measurement, where this is a possible thing if one, at the limit, effects an instantaneous measurement).

Now, for formulating physical laws, which have their only purpose in foreseeing, by means of results which are known at present of measurements which are already completed and which are relative to the past or to the present, the results of other measurements which can be effected in future, the fact of knowing $x(t_0)$ and $p_x(t_0)$ of a same particle at a time $t > t_0$ (where t is the time in which we come to know the result of the measurement of the abscissa x of the particle at the time t_0), in which (time t) the p_x of the particle has already changed in an unknown (to us) way in a whatever quantity, is not useful because this fact does not change our knowledge of the physical state at the present moment, where this knowledge is equal, at the most, to the only, at the limit exact, knowledge of the variable abscissa x of the particle and because we have to base ourselves only on this knowledge of the physical state at the present moment for foreseeing the results of the measurements, which are possible in the future, without or before effecting them.

Therefore, in Quantum Mechanics we construct a theory which is based on the knowledge, at the most, of only one between two physical quantities that are described by operators which do not commute between them (relatively to a same physical system), not because these quantities are absolutely not knowable simultaneously, since we have seen that, in reality, we can know perfectly the values which are assumed simultaneously by these quantities, even though relatively to a past instant, but because for foreseeing the results of measurements, which are completed in the future, by means of results of measurements, which results are known at present and which measurements are completed in the present or in the past, (where these previsions are what interests exclusively in formulating physical laws), we have to start from what is known on the present situation and this situation is known with indeterminacy, even though the situation in a past moment is known with exactness, and therefore the equations of Quantum Mechanics start from physical states which are known approximately for arriving to foresee approximately the future physical states. In fact, in physics, by starting from a partial information on the present we can have, at the most, as prevision an as much partial information on the future.

3 The relativistic case

As for the field of application of Relativistic Quantum Mechanics[2], also in this case we can have free particles with a p_x, which is known with an arbitrarily small, at the limit null, error and which is constant in time, and can measure (at a certain time t_0) the abscissa x of these particles with an arbitrarily small, at the limit null, error, for example by means of a photon of arbitrarily small, at the limit null, wave-length, knowing so both $p_x(t_0)$ and $x(t_0)$ (of every one of these particles) simultaneously with arbitrarily small, at the limit null, errors; but when the measure of the abscissa x of one of these particles with, at the limit, null error comes to our knowledge, not only the p_x of the same particle has changed in a completely unknown way, but particles, which are identical to that of which we measure the abscissa x, are been produced (the identical particles are produced if we measure the abscissa x of the particle with a sufficiently small error) and this takes away from us for all practical purposes, even though not theoretically, also the, at the limit, exact knowledge of the abscissa x of the particle; therefore, for this motive, at the moment in which the measure of the abscissa x of the particle comes to our knowledge, from the point of view of Quantum Mechanics we do not know practically much of useful knowledge on the physical system at the present moment, for which not to use the measures of position is better in Relativistic Quantum Mechanics.

Therefore, in conclusion, except for the production, in certain conditions, of other identical particles, the situation is substantially analogous to that of the non-relativistic case.

---

[2] Cf. BERESTETSKIJ – LIFŠITS – PITAEVSKIJ pp. 15-19.

4 Generalization to any pair of physical quantities

Conducts that are analogous to what we have seen for the $\mathbf{p_x}$ and the $\mathbf{x}$ of a particle can be effected for every pair of physical quantities, which are described by two operators that do not commute between them (relatively to a same physical system) and of which (quantities) one is conserved in time. For example, we have that energy and time are described by operators which do not commute between them (relatively to a same physical system), and that the energy of a physical system is conserved if no (external) force acts on the same physical system.

Moreover, for every pair of physical quantities, which are described by operators that do not commute between them (relatively to a same physical system) and which are not conserved in time between the measurement of one of the two physical quantities and that (the measurement) of the other physical quantity, if we know the exact evolution of one of the two physical quantities in time after that this physical quantity is been measured, the conduct is analogous, because, at the moment of the measurement of the other physical quantity that is described by an operator which does not commute with the operator of the first physical quantity, we know the first physical quantity with an arbitrarily small, at the limit null, error.

While if we do not know the exact evolution in time both of one of the two physical quantities and of the other physical quantity between the two measurements, which we effect, we can always effect them (the two measurements) at an arbitrarily short, at the limit null, interval of time between them, for which, since physical quantities vary with continuity in time, we come to know the first physical quantity with an arbitrarily small, at the limit null, error at the moment of the measurement of the other physical quantity, and therefore we can apply the same reasonings which we have already seen for the **x** and the $\mathbf{p}_x$ of a free particle.

In this way we succeed in lowering without any positive inferior limit, also under the constraint of the inequality of the (Heisenberg's) Uncertainty Principle, the product of the errors of measure on the known values that are assumed simultaneously in the past by two physical quantities, which are described by operators that do not commute between them relatively to a same physical system.

Thus, in conclusion, the situation is similar also here, for which we can say that what we have seen in the non-relativistic case for the **x** and the $\mathbf{p}_x$ of a particle is valid in general.

5 The true meaning of the (Heisenberg's) Uncertainty Principle

We have to underline that the (Heisenberg's) Uncertainty Principle applies in Quantum Mechanics only relatively to the knowledge of the physical state of any physical system, where this knowledge is obtained exclusively by a single measurement, for which these knowledges, which are arbitrarily good relatively to a past moment, do not violate the postulates of Quantum Mechanics in any way, because these knowledges are the result, in reality, of two measurements which are effected one before and the other after. In other words, the (Heisenberg's) Uncertainty Principle is relative only to the knowledge of the physical state of any physical system, which knowledge is useful for foreseeing results of future experiments, for which these knowledges, that are arbitrarily good relatively to a past moment, do not violate the (Heisenberg's) Uncertainty Principle in any way.

In every case we can say that however we have a physical state ψ, which is represented by a (quantum) wave function in Quantum Mechanics, we can in conformity with Quantum Mechanics, at a certain instant according to our choice, measure at the limit without error the physical quantity $\mathbf{A}$ (of the physical state ψ) or the physical quantity $\mathbf{B}$ (of the physical state ψ) even though the operators which are corresponding to these two physical quantities $\mathbf{A}$ and $\mathbf{B}$, (respectively) $\mathbf{A}^{op}$ and $\mathbf{B}^{op}$, do not commute between them, that is to say:

$$[\mathbf{A}^{op}, \mathbf{B}^{op}] \equiv \mathbf{A}^{op}\mathbf{B}^{op} - \mathbf{B}^{op}\mathbf{A}^{op} \neq 0 ,$$

(relatively to a same physical system).

This implies that the physical state ψ has well determinate values at a certain time $\mathbf{t_0}$ both for $\mathbf{A}$, for which the value is equal to $\mathbf{A_\psi}$, and for $\mathbf{B}$, for which the value is equal to $\mathbf{B_\psi}$: because, for example, an, at the limit, exact measure of $\mathbf{A}$ (at a certain time $\mathbf{t_0}$) on the physical state ψ by definition of measurement in Quantum Mechanics does not perturb the measured physical quantity at all during the measurement and because the measure of $\mathbf{A}$, without changes, in conformity with Quantum Mechanics, as a consequence of the measurement at a certain time $\mathbf{t_0}$, results equal to a well determinate value that is to say to $\mathbf{A_\psi}$, we have that the real physical state, in conformity with Quantum Mechanics, had a well determinate value at the time $\mathbf{t_0}$ where this value is equal to $\mathbf{A_\psi}$ also before or without effecting the same measurement. The same, which is valid for $\mathbf{A}$ and $\mathbf{A_\psi}$, is valid also for $\mathbf{B}$ and $\mathbf{B_\psi}$.

We have, therefore, in conformity with Quantum Mechanics, that a quantum wave function, since this wave function contains, for every pair of physical quantities which are described by operators that do not commute between them (relatively to a same physical system), at the most exact information only on one of these two quantities, corresponds to a partial description of a real physical system ψ, which has both its $\mathbf{A}_\psi$ and its $\mathbf{B}_\psi$, that is to say that has well determinate values simultaneously also of physical quantities which are described by operators that do not commute between them (relatively to a same physical system).

Moreover, the indeterminacy of the measures in Quantum Mechanics does not depend, in fact, on the measured object, which has well determinate values, as we have seen, in conformity with Quantum Mechanics, for all measurable physical quantities even though these quantities are described by operators that do not commute between them (relatively to a same physical system); but this indeterminacy depends only on the object (that is used for) measuring, which, for not to modify a measured physical quantity in any way, goes to modify completely the physical quantities that are described by operators which do not commute with the operator of the measured physical quantity (relatively to a same physical system). In particular, this modification is connected to the properties of the elementary particles, which are used as the objects (that are used for) measuring which less perturb the measured object; more precisely, this modification is due to the properties, which are tied to $\hbar$ (that is, to the Planck's constant divided by 2π), of the elementary particles (that are used for) measuring. In fact the $\hbar$, which figures in the (Heisenberg's) Uncertainty Principle, is, in conformity with Quantum Mechanics, relative to the particles (that are used for) measuring, that is to say to the particles that are used as instruments of measurement; it is, likewise, true that also some properties of the measured particles are tied to $\hbar$, since many strictly quantum and microscopic properties are tied to $\hbar$ for all elementary particles both (that are) measured and (that are used for) measuring, but this does not mean in any way, on the contrary of what unfortunately many people have written[3], that the $\hbar$ of the (Heisenberg's) Uncertainty Principle is connected with an intrinsic indeterminacy of the measured object, in fact we have seen that the $\hbar$ of the (Heisenberg's) Uncertainty Principle is not tied to the measured physical system but to the objects that are used for the measurement.

The fact that the $\hbar$, which figures in the (Heisenberg's) Uncertainty Principle, is relative to the perturbative capacities of the particles, which are used as instruments of measurement, on the measured object can be seen clearly also in the limit of Quantum Mechanics for macroscopic measured objects: Classical Mechanics. In this case, that is to say for macroscopic measured objects, the particles, which are used as instruments of measurement, are always the same which are used in the case of Quantum Mechanics for the measurement of microscopic objects and therefore the perturbative capacities of the apparatuses (that are used for) measuring, which use these particles for measuring macroscopic objects, are the same as those in the case of Quantum Mechanics for microscopic measured objects. From this we have that in Classical Mechanics the errors, which are due to the effect of the (Heisenberg's) Uncertainty Principle on the measured physical quantities, are always the same with respect to Quantum Mechanics in absolute value and therefore practically negligible in relative value compared to macroscopic physical quantities, which are observed in Classical Mechanics. Therefore, we have that, contrary to what many people unfortunately assert[4], the indeterminacy, which is due to the

---

[3] Cf. LANDAU – LIFSHITZ pp. 15-21; BERESTETSKIJ – LIFŠITS – PITAEVSKIJ pp. 15-19; HAWKING pp. 72-77; EISBERG – RESNICK pp. 65-68.77-80; BRANSDEN – JOACHAIN p. 58.

[4] Cf. LANDAU – LIFSHITZ pp. 15-21; BERESTETSKIJ – LIFŠITS – PITAEVSKIJ pp. 15-19; HAWKING pp. 72-77; EISBERG – RESNICK pp. 65-68.77-80; BRANSDEN – JOACHAIN p. 58.

(Heisenberg's) Uncertainty Principle, is not intrinsic to the measured physical system: in fact, if the indeterminacy were intrinsic to the measured physical system, we would have had, increasing the number N of the particles of the measured physical system, for the physical quantities, which are expressed by the sum of the contributions of every single particle, as for example the momentum p, an error that is increasing in a directly proportional way with N (since by definition of error we have, universally in physics, that the error for a sum of physical quantities is equal to the sum of the errors for the single physical quantities), while for the physical quantities, which are not given by the sum of the contributions of the N particles, as for example the abscissa x (where x is understood as the abscissa of a point, for example of a rim, of the physical system) an error which is not increasing but constant as N increases; for which

$$\Delta p_x \cdot \Delta x \geq N \frac{\hbar}{2}$$

would be the (Heisenberg's) Uncertainty Principle for x and p_x in the classical case according to this incorrect interpretation of the (Heisenberg's) Uncertainty Principle. Obviously, we would have analogous things for p_y and y, and for p_z and z. Instead, in the case of energy and time, because of the fact that the energy E of the physical system increases as N increases and that the time t of the physical system does not increase but is constant as N increases, by means of a similar reasoning we would have that

$$\Delta E \cdot \Delta t \geq N \frac{\hbar}{2}$$

would be the (Heisenberg's) Uncertainty Principle for energy and time in the classical case according to this incorrect interpretation of the (Heisenberg's) Uncertainty Principle. Now, these expressions of the (Heisenberg's) Uncertainty Principle in the classical case are clearly false, because we can, in conformity with the postulates of Quantum Mechanics, independently of the size of the measured object and of how many elementary particles constitute the measured object, always measure two physical quantities, that are described by operators which do not commute between them (relatively to a same physical system), by means of a single operation of measurement, with products of errors which are small down to a theoretically attainable least of $\frac{\hbar}{2}$ and not of $N\frac{\hbar}{2}$, that is to say with products of errors which, for the case of $\mathbf{p_x}$ and $\mathbf{x}$, are such that

$$\Delta\mathbf{p_x} \cdot \Delta\mathbf{x} \geq \frac{\hbar}{2} \; ;$$

and, for the case of energy and time, are such that

$$\Delta\mathbf{E} \cdot \Delta\mathbf{t} \geq \frac{\hbar}{2} \; .$$

Therefore, we have that this reasoning gives us an ulterior proof of the falsity of this incorrect interpretation of the (Heisenberg's) Uncertainty Principle, from which interpretation this same reasoning was started.

Moreover, we have, as a consequence of these considerations, that the errors, which are not eliminable according to the (Heisenberg's) Uncertainty Principle do not depend, in accordance with Quantum Mechanics, in any way on the physical system, which is under observation and which, therefore, intrinsically, in accordance with Quantum Mechanics, has well definite simultaneously without error also the values of physical quantities that are described by operators which do not commute between them (relatively to a same physical system).

Therefore we cannot say that the only part physically observable of a physical state, in accordance with Quantum Mechanics, is given by a quantum wave function because, as we have seen, effecting the measurements we go to measure, from the point of view of Quantum Mechanics, on a real physical state which has well determinate simultaneously also the values of two physical quantities which are described by two corresponding operators that do not commute between them (relatively to a same physical system), while the quantum wave function has not both of these values determinate, because it has a part of information less than the real physical state, in accordance with Quantum Mechanics. In other words, we can, in accordance with Quantum Mechanics, at the limit without error, measure both the physical quantity $\mathbf{A}$ and the physical quantity $\mathbf{B}$ on a real physical state, even though the corresponding operators $\mathbf{A^{op}}$ and $\mathbf{B^{op}}$ do not commute between them, that is to say

$$[\mathbf{A^{op}}, \mathbf{B^{op}}] \equiv \mathbf{A^{op}B^{op}} - \mathbf{B^{op}A^{op}} \neq 0,$$

(relatively to a same physical system). Therefore the real physical state $\mathbf{\psi}$, according to Quantum Mechanics, has both a well determinate value for the physical quantity $\mathbf{A}$, where this value is equal to $\mathbf{A_{\psi}}$, and a well determinate value for the physical quantity $\mathbf{B}$, where this value is equal to $\mathbf{B_{\psi}}$, even though before measurement we do not know both these values and what we know before measurement is given by a quantum wave function.

According to Quantum Mechanics, we have that $\mathbf{A}_\psi$ and $\mathbf{B}_\psi$, being measured physically at the limit without error, are physical quantities from every point of view; therefore to say that a real physical state is completely described from every physical point of view by a quantum wave function, which, in reality, according to Quantum Mechanics, contains only a partial information of the real physical state, is equivalent to not to consider a part of the physical world, which part is subject to physical measurements, as object of physics. For example, to consider real, from the physical point of view, only a physical quantity, of which the quantum wave function is an eigenstate, between two physical quantities, which are described by two corresponding operators that are not commuting between them (relatively to a same physical system), is equivalent to arbitrarily not to consider the other physical quantity, which is described by an operator that does not commute with the operator of the former (relatively to a same physical system), as real from the physical point of view, but this other physical quantity is instead real from every physical point of view, because this other physical quantity is really physically measurable, at the limit without error, in accordance with Quantum Mechanics.

Therefore we cannot, according to Quantum Mechanics, limit ourselves to the quantum wave function as expression of the real physical state, but we have to look at the quantum wave function as the partial knowledge which we have on the real physical state, where this partiality of knowledge is due to the errors of measure.

On the other hand, the errors of measure, more or less large, are always been present in physics and here in Quantum Mechanics we do not have anything of new as for the presence of the errors of measure; but here in Quantum Mechanics we use a new method for limiting the damages, which are due to the errors of measure.

In conclusion, all the attempts, which unfortunately many people make, of denying the physical realities, which are been reaffirmed in this treatment, knock against the postulates of Quantum Mechanics. In particular, to attribute changes, which are strange and absurd from the physical point of view, of a physical quantity to the operations of measurement of the same physical quantity is incorrect and without sense, in fact this is not in any way justified, in conformity with Quantum Mechanics, because there are postulates of Quantum Mechanics which assert that one can measure a whatever physical quantity, at the limit without error, without modifying this same physical quantity in any way during the measurement.

6 The correct interpretation of the famous double-slit experiment

In the famous double-slit experiment[5] the figure of interference is present even though the intensity of the source is so weak that the particles (for example electrons) are seen on the screen one at a time.

Now, in this experiment when there are many particles, they (these particles) interferes between them. Obviously, this interference is due to the properties of the measured particles, which properties are different from the classical properties at the level of magnitude $\hbar$ (that is, the Planck's constant divided by 2π). On the other hand, the indeterminacy is relatively great for a single particle but not for the macroscopic ensemble of all the particles, for which ensemble the indeterminacy is relatively negligible.

Instead, in the case that the particles (for example electrons) are seen on the screen one at a time, we know only that the

---

[5] Cf. GASIOROWICZ pp. 20-21.34-36; HAWKING pp. 75-77.

intensity of the source is very weak and that the structure of the experiment is always the same of the case with the macroscopic ensemble of all the particles: therefore, every time that we see something on the screen is as if we had taken a single particle from the macroscopic ensemble of all the particles, in virtue of the fact that is as if we were emitting the correspondent macroscopic ensemble of all the particles a particle at a time. So it is not a surprise that we have also in this case a figure of interference on the screen. On the other hand, in this case to reason about which split the particle that is seen on the screen has gone through is erroneous and very misleading: in fact, we are seeing only the macroscopic ensemble of all the particles a particle at a time!

Moreover, obviously, a modification of the experimental conditions can modify the macroscopic ensemble of all the particles: for example, if we close one of the two slits, the macroscopic ensemble of all the particles is modified, so that the figure of interference on the screen is no longer present. But this is not mysterious in any way, because it is clear that the experimental conditions have to influence the macroscopic ensemble of all the particles.

Thus, in conclusion, the double-slit experiment is, in reality, entirely not mysterious and not paradoxical.

7 The correct interpretation of the Einstein-Podolsky-Rosen Paradox

Moreover, we have that some phenomena, as that of the Einstein-Podolsky-Rosen Paradox[6], result as paradoxes only by means of incorrect reasonings, which are unfortunately effected by many books. In fact, in reality, reasoning correctly according

---

[6] Cf. EINSTEIN – PODOLSKY – ROSEN pp. 777-780; SAKURAI pp. 225-227.

to Quantum Mechanics, we have that, for example, there is not anything of paradoxical in the Einstein-Podolsky-Rosen Paradox. On the contrary, in reality, if the Einstein-Podolsky-Rosen Paradox had not been true we would have had a completely paradoxical and absurd phenomenon; in fact, in reality, according to Quantum Mechanics, in the phenomenon of the Einstein-Podolsky-Rosen Paradox when we measure (in the case of this Paradox with a couple of photons) on a photon (of the couple) what changes, relatively to the other photon, is the quantum wave function of the other photon, where this wave function represents our knowledge of the real physical state of the other photon, but the real physical state of the other photon does not change and remains unaltered, for which there is not anything of paradoxical, according to the correct interpretation of Quantum Mechanics, in the phenomenon of the Einstein-Podolsky-Rosen Paradox, because by means of measuring on a photon (of the couple) we, by definition of measurement in Quantum Mechanics, have not changed in any way the other photon which maintains the same real physical state.

This (phenomenon) and many other phenomena appear as "incurable" paradoxes only for the fact of wanting to consider, in opposition to the postulates of Quantum Mechanics, the quantum wave function of a physical system as the real physical state of the same physical system from every physical point of view and for the fact of wanting to consider, in opposition to the postulates of Quantum Mechanics, two physical quantities, which are described by two corresponding operators that do not commute between them (relatively to a same physical system), as physical quantities that do not exist simultaneously relatively to a same physical system.

The fact that several absurd and incurable paradoxes come out from these incorrect interpretations of Quantum Mechanics is, obviously, an ulterior proof of the errors, from the physical point of view, that are inherent in these interpretations of

Quantum Mechanics, which interpretations are, on the other hand, contrary, as we have seen, to the postulates of Quantum Mechanics.

8 The correct interpretation of the Bell's Theorem

As for the Bell's Theorem[7] the instantaneous correlations between the particles are relative only to our knowledge about them: as we have seen in the case of the Einstein-Podolsky-Rosen Paradox, the apparent change in a particle is not, in reality, due to an interaction with another particle that is measured and that is correlated with the first particle, but only to the change in our knowledge about the first particle.

In other words, this theorem does not demonstrate the presence of an instantaneous (and therefore superluminal) interaction between the measured particle and another particle that is correlated with the first particle, but (demonstrates) only the instantaneous (and therefore, in a certain sense, superluminal) interaction between our knowledge of the first particle and our knowledge of another particle that is correlated with the first particle.

In conclusion, the Bell's Theorem is not in conflict with my position, but only in conflict (for the problematic alleged presence of superluminal and mysterious interactions between the two measured particles) with other interpretations of Quantum Mechanics, which interpretations we have already confuted above. Therefore this theorem is not in opposition to but is in support of my interpretation of Quantum Mechanics.

---

[7] Cf. GHIRARDI pp. 203-233; FISCALETTI pp. 77-89.

9 Conclusion: Quantum Mechanics is incomplete

Therefore, in conclusion, we can say that Quantum Mechanics is incomplete, since the quantum wave function of a physical system does not describe completely the real physical state of the same physical system. In fact we have seen that in general two physical quantities of a physical system, which are described by operators that do not commute between them (relatively to a same physical system), have really determinate values that are not contained in the quantum wave function of the same physical system; therefore, the description of any physical system by means of a quantum wave function is an incomplete description of the same physical system, and consequently Quantum Mechanics is incomplete.

CHAPTER II
THE SECOND QUANTIZATION:
THE CORRECT INTERPRETATION
OF THE FEYNMAN DIAGRAMS

This correct interpretation of the (Heisenberg's) Uncertainty Principle must be applied also to the second quantization, in particular to the interpretation of the Feynman diagrams. Therefore, we must say that also in the phenomena that are described by these diagrams there is always the conservation of energy, also below the value of ΔE in the formula of the (Heisenberg's) Uncertainty Principle:

$$\Delta E \cdot \Delta t \geq \frac{\hbar}{2}.$$

In other words, it is not possible that the energy not be conserved for a time Δt by a quantity $\leq \Delta E$, even though this value is less than (or equal to) the indeterminacy of the energy in the above formula of the (Heisenberg's) Uncertainty Principle.

On the other hand, it is true that in the central area of these diagrams the energy is not always conserved in every vertex, but these vertexes in which the energy is not conserved cannot be present alone but only with other vertexes (or with another vertex) so that the energy is conserved on the whole: in other words, we must consider these vertexes together since a vertex cannot be present alone if in it the energy is not conserved. We have to make so because the effective phenomena are not the single vertexes of these diagrams but only the groups of vertexes in which the energy is conserved: in other words, the single vertexes of the Feynman diagrams are a simplification that is not always present in reality. Moreover, it is not possible to say that a vertex in which there is not the conservation of energy happens alone previously and then the other vertexes that compensate (or the other vertex that compensates) the violation of the conservation of energy happen (or happens), since the first vertex cannot happen in any case if the other vertexes do not happen (or the other vertex does not happen): in other words, the effective phenomenon is the entire group of vertexes that on the whole conserve the energy.

More in general the validity of the (Heisenberg's) Uncertainty Principle does not permit us to say that the reality is intrinsically indeterminate with regard to the application of the physical laws, because the same physical laws apply to all physical quantities, also below the uncertainties of the (Heisenberg's) Uncertainty Principle, even though in this case we do not know those physical quantities with infinite accuracy.

In conclusion, both the microscopic physical realities and the physical laws that apply on the microscopic realities are completely determinate, even if the (Heisenberg's) Uncertainty Principle is valid, since the correct interpretation of this principle shows clearly that this principle does not limit in any way the determinateness of the microscopic physical realities and of the physical laws.

CHAPTER III
THE INCOMPLETENESS OF QUANTUM MECHANICS IS ALSO PROVED BY THE POSSIBILITY OF USING THE THEORY OF DE BROGLIE-BOHM

The possibility (which is widely recognized) of using the Theory of De Broglie-Bohm for the quantum phenomena[8] proves that Quantum Mechanics does not imply that the reality is intrinsically indeterminate, since this theory is a deterministic theory even though it is not local. In other words, it is not Quantum Mechanics to say that the reality is intrinsically indeterminate but only the traditional interpretation of Quantum Mechanics, that is to say the Copenhagen Interpretation, and other similar interpretations (of Quantum Mechanics).

On the other hand, the Theory of De Broglie-Bohm even though is deterministic is not local, that is to say this theory contemplates instantaneous interactions between the particles, and these interactions are completely strange and mysterious. Because of this, my position is superior also to the Theory of De Broglie-Bohm, since my position asserts the determinism in the field of quantum physics without introducing superluminal and mysterious interactions.

In other words, the Theory of De Broglie-Bohm is a good theory only if we interpret the not local interaction of this theory

---

[8] Cf. FISCALETTI pp.179-260; C. CALLENDER in CRAIG – SMITH pp. 50-72; Q. SMITH in CRAIG – SMITH pp. 73-124; A. VALENTINI in CRAIG – SMITH pp. 125-155; T. MAUDLIN in CRAIG – SMITH pp. 156-179; CRAIG pp. 142.227-229.

as relative only to our knowledge of the physical state of the physical system, which knowledge can change instantaneously when we effect measurements or when we change the experimental conditions. We have a good application of this in the case of my explanation of the famous double-slit experiment: when we close one of the two slits we are changing the experimental conditions and therefore we are changing the composition of the macroscopic ensemble of all the particles, from which we take a particle at a time. So it is clear that it is not necessary, contrary to the Theory of De Broglie-Bohm, any mysterious instantaneous interaction between the particles for explaining the experimental results!

On the other hand, we could say that my interpretation of Quantum Mechanics is an improvement of the Theory of De Broglie-Bohm, since my interpretation is similar to that of this theory except for the fact that I have replaced the mysterious and simultaneous interactions of this theory with the variations of our knowledge of the physical systems.

GENERAL CONCLUSION: QUANTUM MECHANICS IS INCOMPLETE AND NOT PARADOXICAL

Therefore, we have seen that many paradoxes arise in Quantum Mechanics only for the incorrect interpretation of the (effective and thought) experimental results and of the (Heisenberg's) Uncertainty Principle. Once we have corrected these erroneous interpretations, we see clearly that the Copenhagen Interpretation and the other analogous Interpretations (of Quantum Mechanics) are untenable, and that Quantum Mechanics, even though diverse obviously from Classical Mechanics, is incomplete and entirely not paradoxical.

Moreover, we have seen that once we interpret correctly Quantum Mechanics (improving the Theory of De Broglie-Bohm by means of the replacement of the mysterious and simultaneous interactions of this theory with the variations of our knowledge of the physical systems), Quantum Mechanics is entirely not paradoxical. On the other hand, this correct interpretation constitutes my interpretation of Quantum Mechanics and is clearly superior to all the other contemporary interpretations of Quantum Mechanics. In other words, my interpretation is similar to that of the Theory of De Broglie-Bohm except for the fact that I have replaced the mysterious and simultaneous interactions of this theory with the variations of our knowledge of the physical systems.

In conclusion, the principle that everything that is not intrinsically necessary has a cause is valid also in the field of application of Quantum Mechanics. Moreover, the physical laws are valid also on all the microscopic realities, and all the microscopic realities have intrinsically determinate values. On the other hand, there are some properties of the particles, which properties we see only in the field of application of Quantum Mechanics: these properties are connected to the Planck's constant **h**, in particular to the fact that this constant is not zero.

BIBLIOGRAPHY

BERESTETSKIJ V. B. – LIFŠITS E. M. – PITAEVSKIJ L. P., *Teoria quantistica relativistica*, Editori Riuniti, Roma 1978 {Italian translation (with some variations made by the same authors) of BERESTETSKII V. B. – LIFSHITZ E. M. – PITAEVSKII L. P., *Reljativistskaja kvantovaja teorija*, Nauka, Moskow 1968-1971}.

BRANSDEN B. H. – JOACHAIN C. J., *Physics of Atoms and Molecules*, Longman Scientific & Technical, Harlow (Essex UK) 1983.

CRAIG W. L., *Time and the Metaphysics of Relativity*, Kluwer Academic Publishers, Dordrecht 2001.

CRAIG W. L. – SMITH Q. (eds.), *Einstein, Relativity and Absolute Simultaneity*, Routledge, London 2007.

EINSTEIN A. – PODOLSKY B. – ROSEN N., *Can Quantum-Mechanical Description of Physical Reality Be Considered Complete?*, Physical Review 47, (1935) 777-780.

EISBERG R. – RESNICK R., *Quantum Physics of Atoms, Molecules, Solids, Nuclei, and Particles*, John Wiley & Sons, New York 1985^2.

FISCALETTI D., *I gatti di Schrödinger*, Editori Riuniti university press, Roma 2015.

GASIOROWICZ S., *Quantum Physics*, John Wiley & Sons, New York 1974.

GHIRARDI G. C., *Un'occhiata alle carte di Dio*, il Saggiatore (Net), Milano 2003.

HAWKING S., *Dal Big Bang ai Buchi Neri*, Biblioteca Universale Rizzoli, Milano 1997 {Italian translation of HAWKING S., *A Brief History of Time*, Bantam Dell Publishing Group, New York 1988}.

LANDAU L. D. – LIFSHITZ E. M., *Meccanica quantistica: Teoria non relativistica*, Editori Riuniti, Roma 1982 {Italian translation of LANDAU L. D. – LIFSHITZ E. M., *Kvantovaja mechanica*, Nauka, Moscow 1973[3]}.

SAKURAI J. J., *Advanced Quantum Mechanics*, Addison-Wesley Publishing Company, Reading (Massachusetts) 1987[11].

TABLE OF CONTENTS

Finito di stampare nel mese di Luglio 2016
per conto di Youcanprint *Self-Publishing*